Things Move

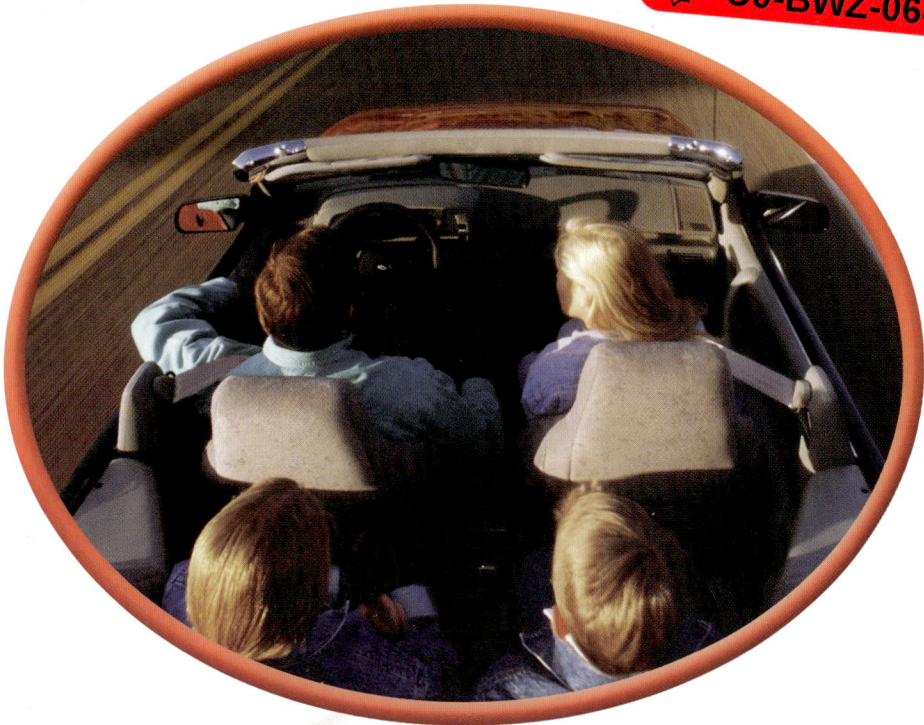

by Betsy Lewis and Cynthia Swain

A bike can go.

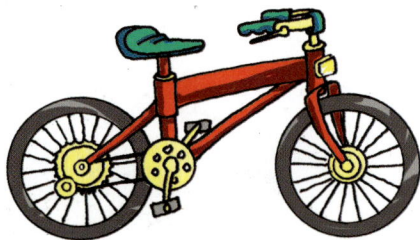

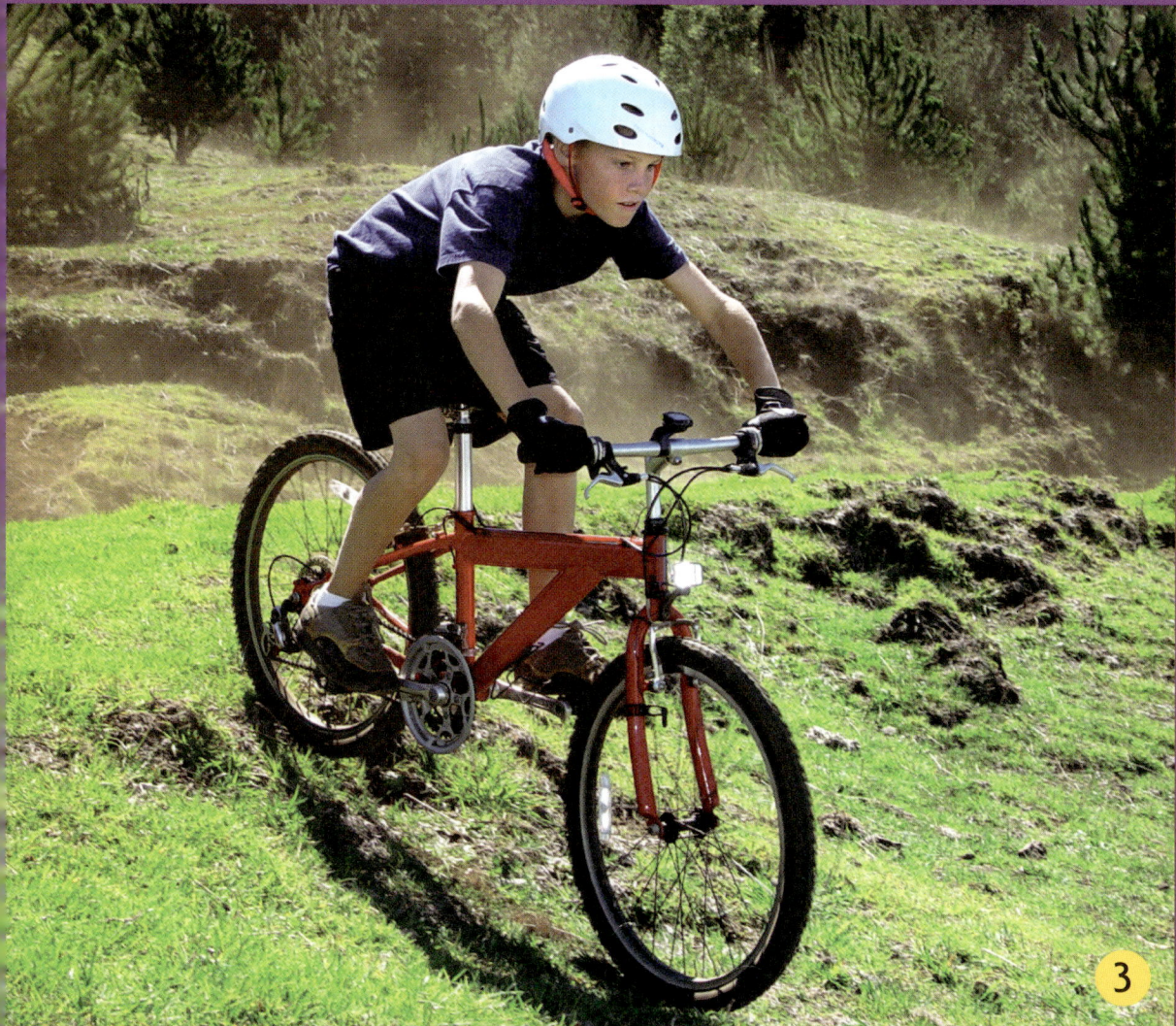

A scooter can go.

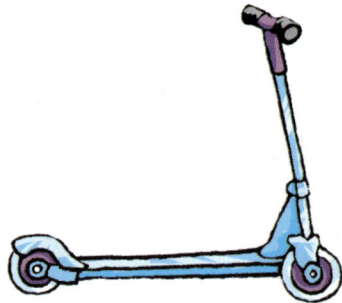

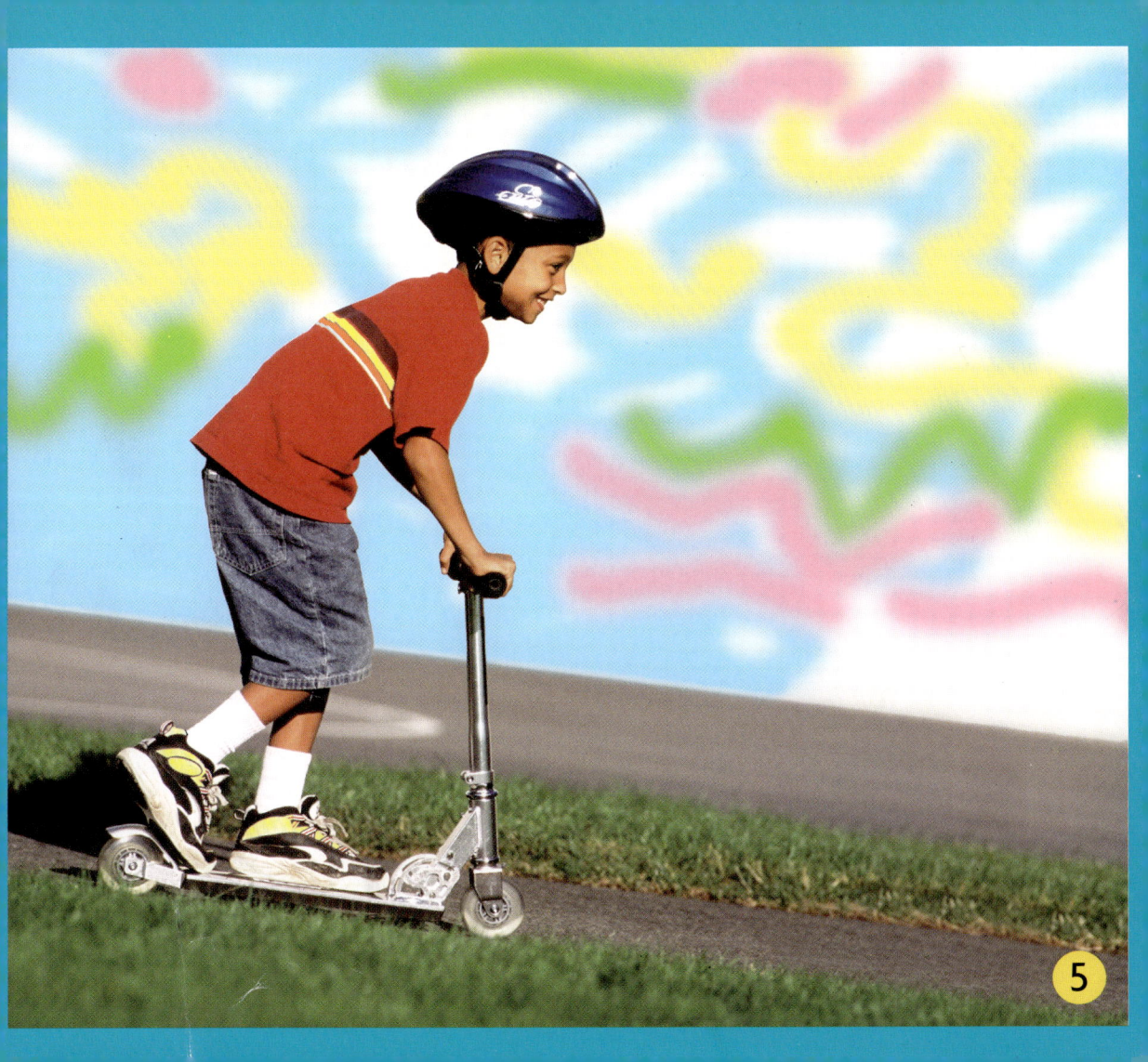

A sled can go.

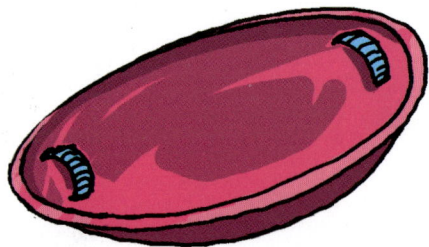

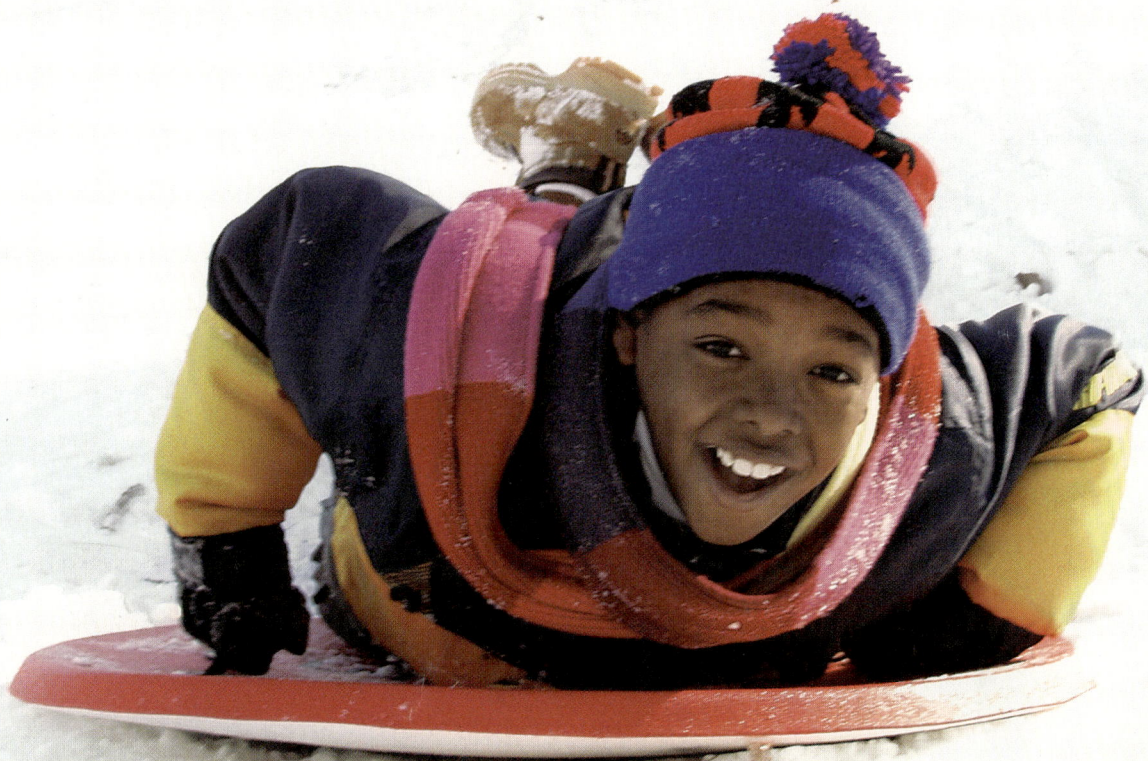

A car can go.

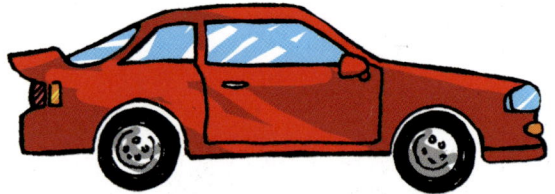

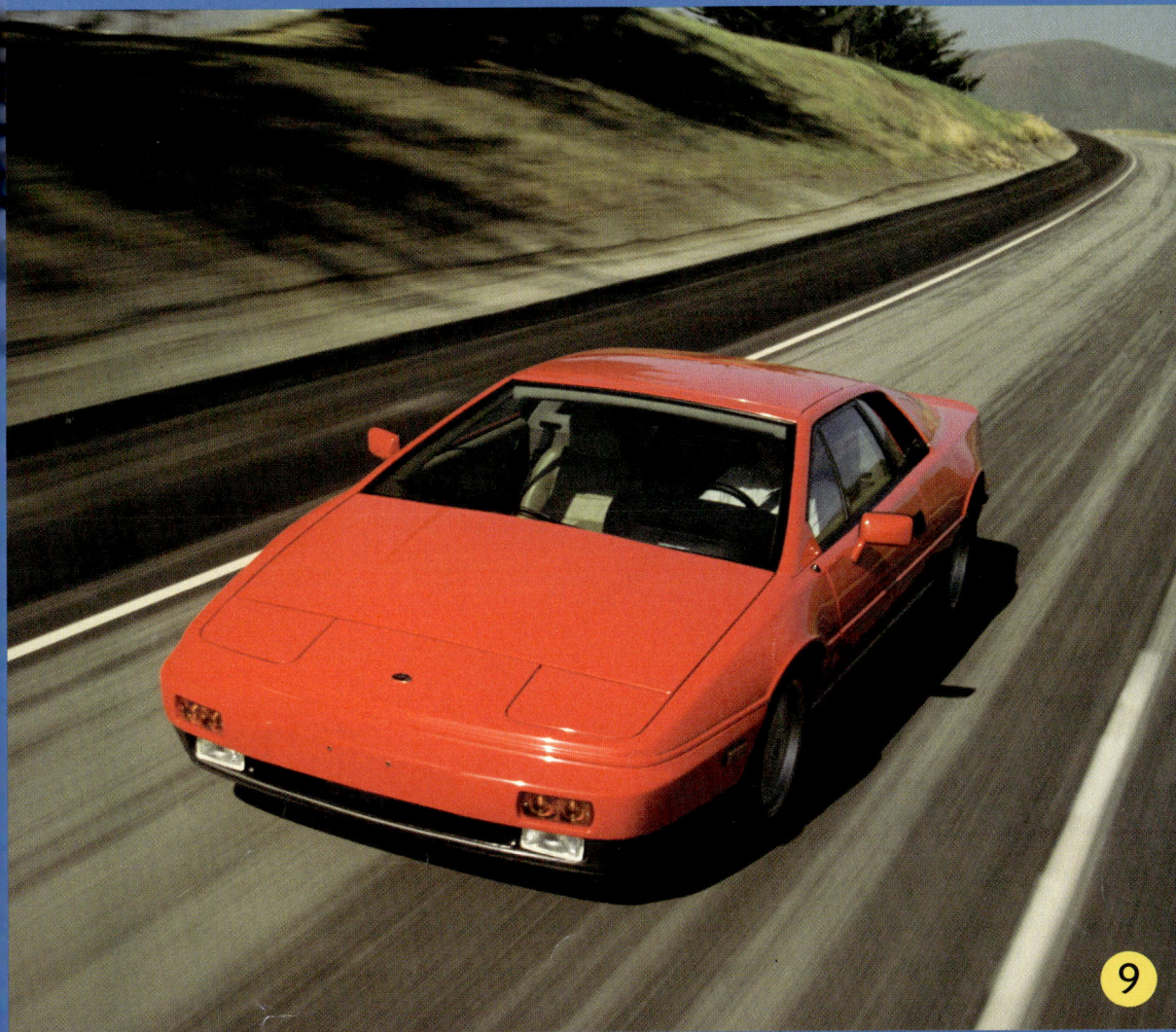

A submarine can go.

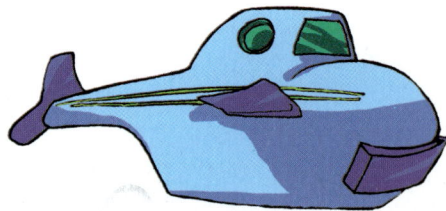

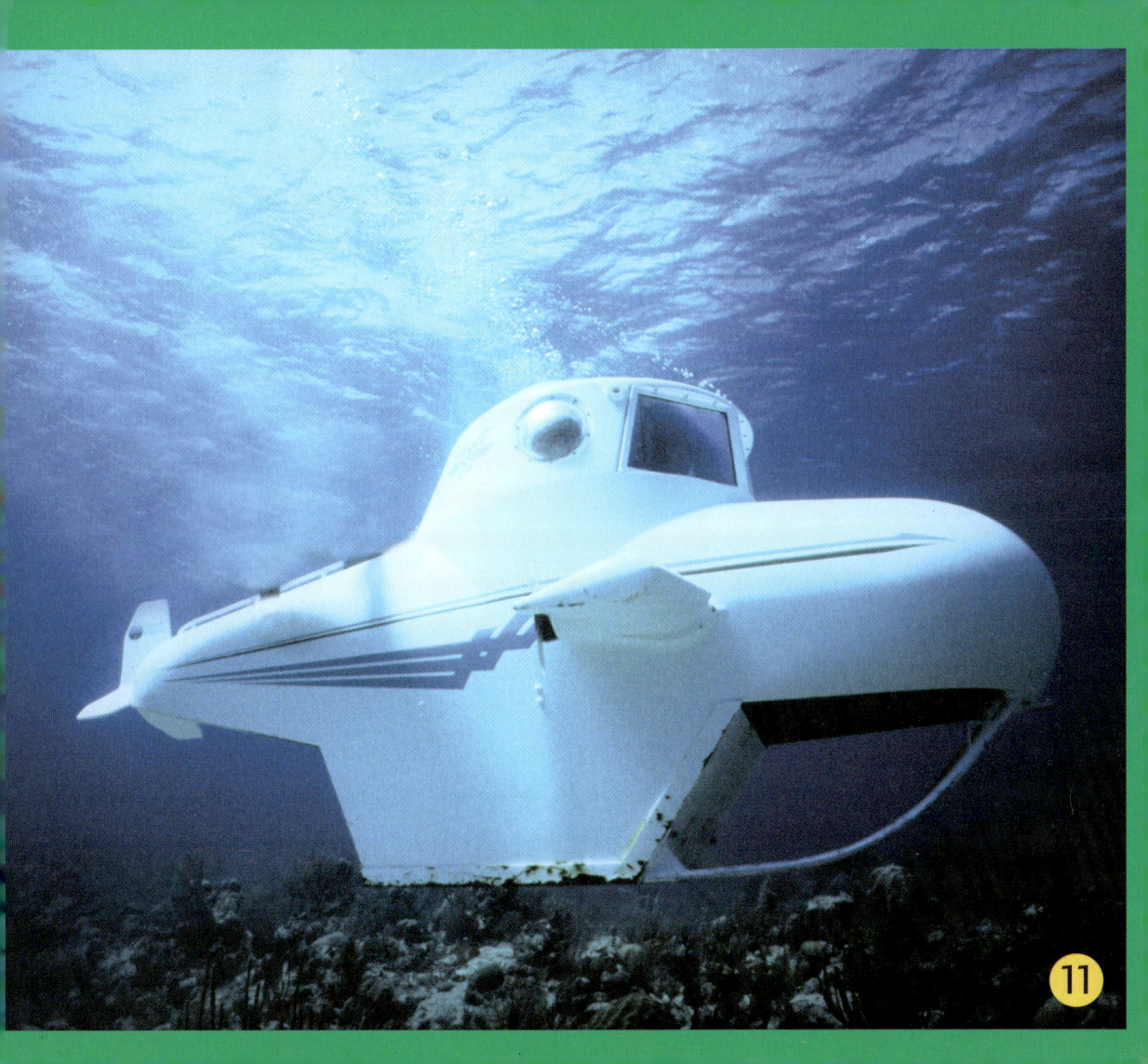

11

A motorcycle can go.

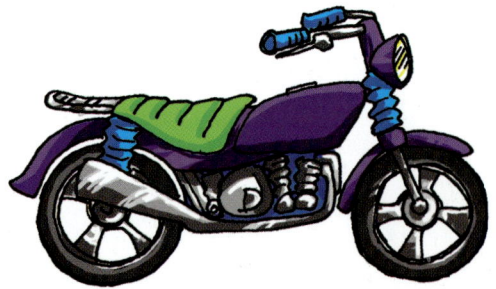

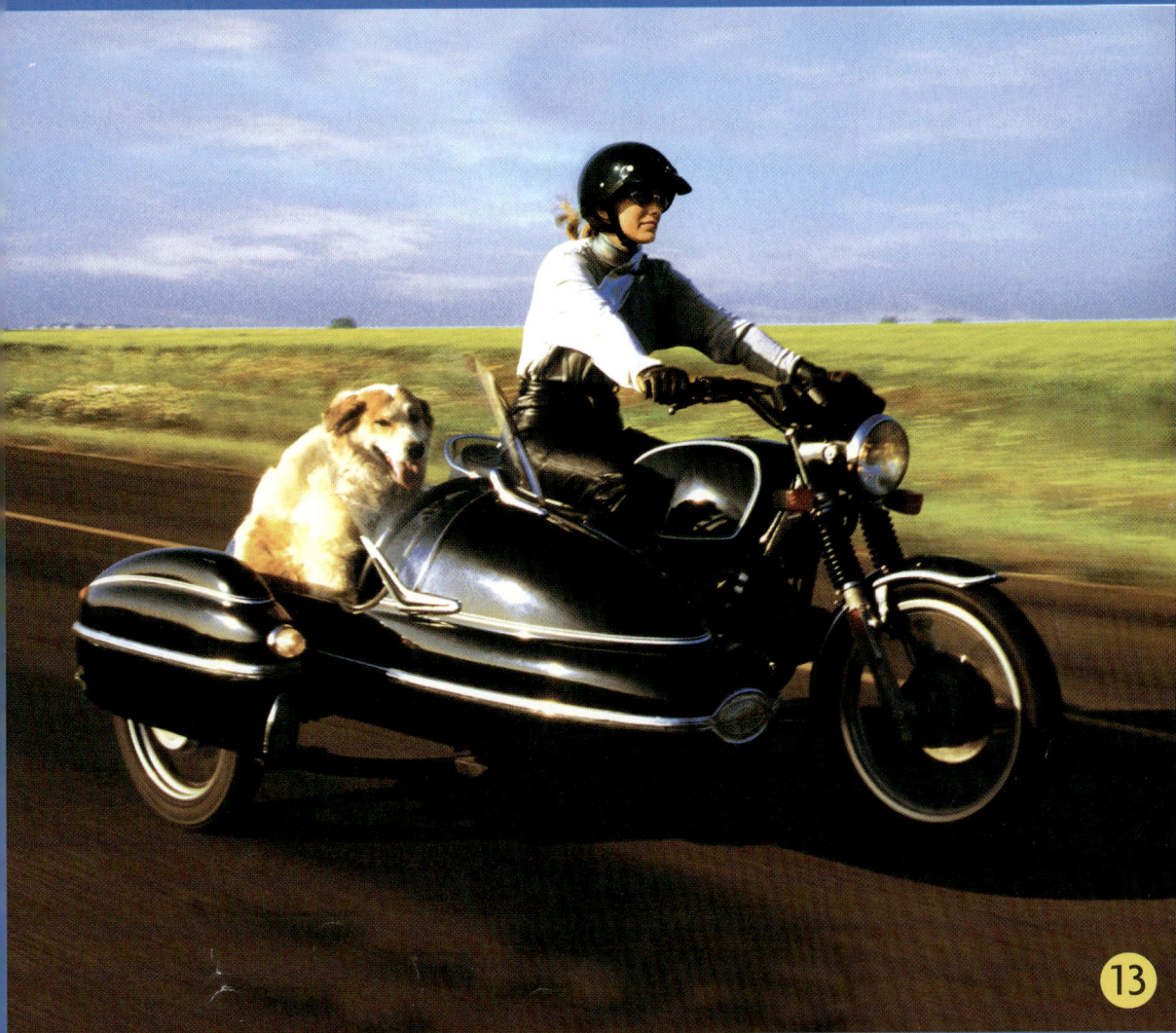

A helicopter can go.

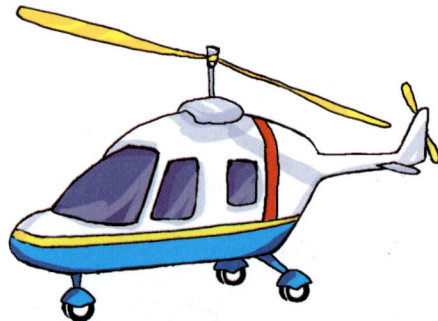

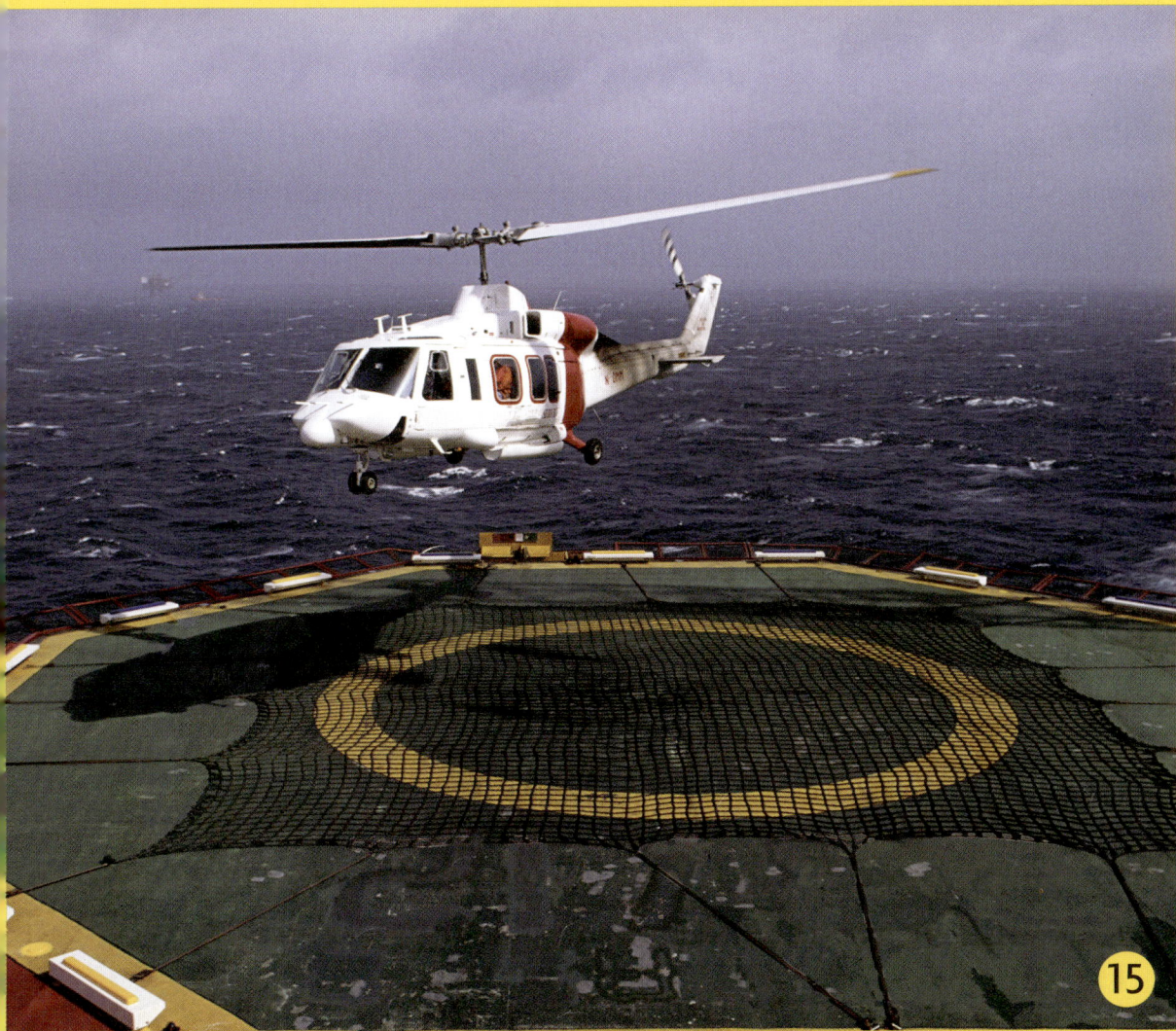

A roller coaster can go!

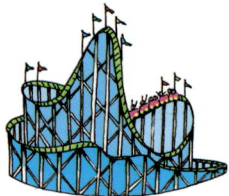

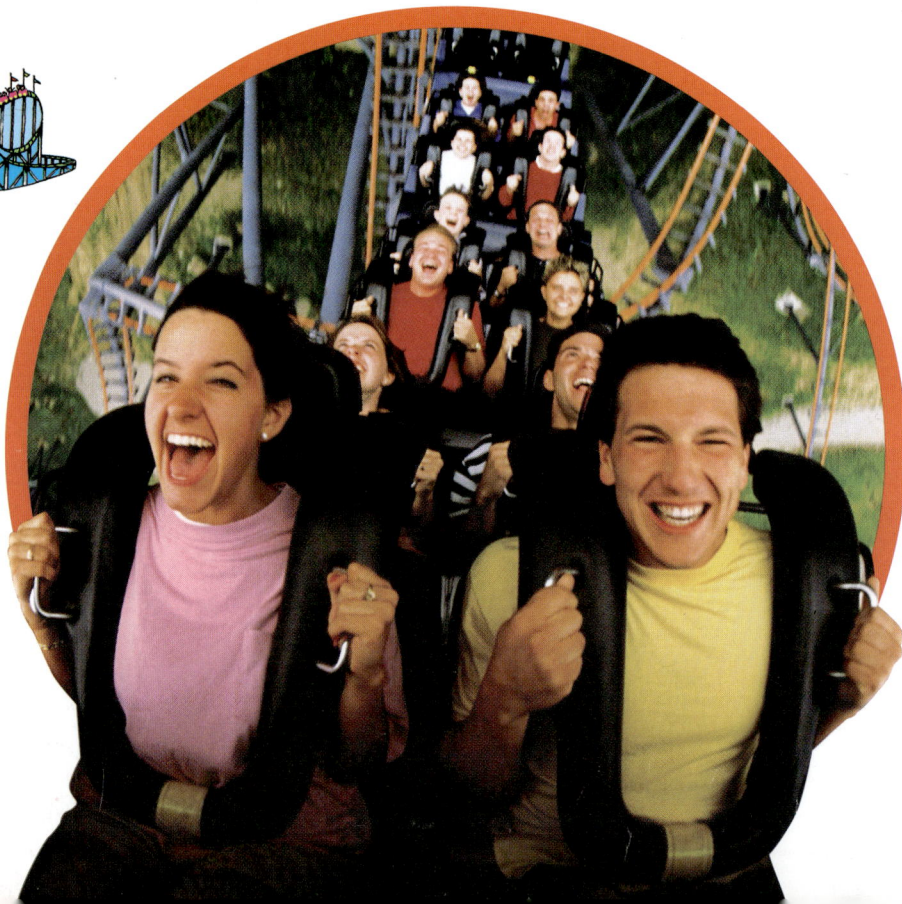